Leckie × Leckie
Scotland's leading educational publishers

D0178885

Practice Papers for SQA Exams

National 5

Maths

001/26012014

10 9 8 7 6 5 4 3

ISBN 9780007504718

Published by
Leckie & Leckie Ltd
An imprint of HarperCollins*Publishers*
Westerhill Road, Bishopbriggs, Glasgow, G64 2QT
T: 0844 576 8126 F: 0844 576 8131
leckieandleckie@harpercollins.co.uk
www.leckieandleckie.co.uk

Special thanks to
QBS (layout and illustration); Ink Tank (cover design); Roda Morrison (copy-edit); Paul Sensecall (proofread); Laura Clegg (proofread)

A CIP Catalogue record for this book is available from the British Library.

Acknowledgements
Whilst every effort has been made to trace the copyright holders, in cases where this has been unsuccessful, or if any have inadvertently been overlooked, the Publishers would gladly receive any information enabling them to rectify any error or omission at the first opportunity.

Printed in Great Britain by Martins the Printers

Introduction

Layout of the Book

This book contains practice exam papers, which mirror the actual SQA exam as much as possible. The layout, paper colour and question level are all similar to the actual exam that you will sit, so that you are familiar with what the exam paper will look like.

The solutions section is at the back of the book. The full worked solution is given to each question so that you can see how the right answer has been arrived at. The solutions are accompanied by a commentary which includes further explanations and advice. There is also an indication of how the marks are allocated and, where relevant, what the examiners will be looking for. Reference is made at times to the relevant sections in Leckie & Leckie's book Success Guide: National 5 Mathematics

Revision advice is provided in this introductory section of the book, so please read on!

How To Use This Book

The Practice Papers can be used in two main ways:

1. You can complete an entire practice paper as preparation for the final exam. If you would like to use the book in this way, you might want to complete each practice paper under exam style conditions by setting yourself a time for each paper and answering it as well as possible without using any references or notes. Alternatively, you can answer the practice paper questions as a revision exercise, using your notes to produce a model answer. Your teacher may mark these for you.

2. You can use the Topic Index at the front of this book to find all the questions within the book that deal with a specific topic. This allows you to focus specifically on areas that you particularly want to revise or, if you are mid-way through your course, it lets you practise answering exam-style questions for just those topics that you have studied.

Revision Advice

Work out a revision timetable for each week's work in advance – remember to cover all of your subjects and to leave time for homework and breaks. For example:

Day	**6pm–6.45pm**	**7pm–8pm**	**8.15pm–9pm**	**9.15pm–10pm**
Monday	Homework	Homework	English Revision	Chemistry Revision
Tuesday	Maths Revision	Physics revision	Homework	Free
Wednesday	Geography Revision	Modern Studies Revision	English Revision	French Revision
Thursday	Homework	Maths Revision	Chemistry Revision	Free
Friday	Geography Revision	French Revision	Free	Free
Saturday	Free	Free	Free	Free
Sunday	Modern Studies Revision	Maths Revision	Modern Studies Revision	Homework

Make sure that you have at least one evening free a week to relax, socialise and re-charge your batteries. It also gives your brain a chance to process the information that you have been feeding it all week.

Arrange your study time into one hour or 30 minute sessions, with a break between sessions, e.g. 6pm –7pm, 7.15pm –7.45pm, 8pm–9pm. Try to start studying as early as possible in the evening when your brain is still alert and be aware that the longer you put off starting, the harder it will be to start!

Study a different subject in each session, except for the day before an exam.

Do something different during your breaks between study sessions – have a cup of tea, or listen to some music. Don't let your 15 minutes expand into 20 or 25 minutes though!

Have your class notes and any textbooks available for your revision to hand as well as plenty of blank paper, a pen, etc. You should take note of any topic area that you are having particular difficulty with, as and when the difficulty arises. Revisit that question later having revised that topic area by attempting some further questions from the exercises in your textbook.

Revising for a Maths exam is different from revising for some of your other subjects. Revision is only effective if you are trying to solve problems. You may like to make a list of 'Key Questions' with the dates of your various attempts (successful or not!). These should be questions that you have had real difficulty with.

Key Question	**1st Attempt**		**2nd Attempt**		**3rd Attempt**	
Textbook P56 Q3a	18/2/14	×	21/2/14	✔	28/2/14	✔
Practice Exam A Paper 1 Q5	25/2/14	×	28/2/14	×	3/3/14	
2014 SQA Paper, Paper 2 Q4c	27/2/14	×	2/3/14			

The method for working this list is as follows:

1. Any attempt at a question should be dated.

2. A tick or cross should be entered to mark the success or failure of each attempt.

3. A date for your next attempt at that question should be entered:
for an unsuccessful attempt – 3 days later
for a successful attempt – 1 week later

4. After two successful attempts remove that question from the list (you can assume the question has been learnt!)

Using 'The List' method for revising for your Maths Exam ensures that your revision is focused on the difficulties you have had and that you are actively trying to overcome these difficulties.

Finally forget or ignore all or some of the advice in this section if you are happy with your present way of studying. Everyone revises differently, so find a way that works for you!

Transfer Your Knowledge

As well as using your class notes and textbooks to revise, these practice papers will also be a useful revision tool as they will help you to get used to answering exam style questions. As you work through the questions you may find an example that you haven't come across before. Don't worry! There may be several reasons for this. The question may be on a topic that you have not yet covered in class. Check with your teacher to find out if this is the case. Or it could be that the wording or the context of the question is unfamiliar. This often happens with reasoning questions in the Maths exam. Once you have familiarised yourself with the worked solutions, in most cases you will find that the question is using mathematical techniques which you are familiar with. In either case you should revisit that question later to check that you can successfully solve it.

Trigger Words

In the practice papers and in the exam itself, a number of 'trigger words' will be used in the questions. These trigger words should help you identify a process or a technique that is expected in your solution to that part of the question. If you familiarise yourself with these trigger words, it will help you to structure your solutions more effectively.

Trigger Words	**Meaning/Explanation**
Evaluate	Carry out a calculation to give an answer that is a value.
Hence	You must use the result of the previous part of the question to complete your solution. No marks will be given if you use an alternative method that does not use the previous answer.
Simplify	This means different things in different contexts: Surds: reduce the number under the root sign to the smallest possible by removing square factors. Fractions: one fraction, cancelled down, is expected. Algebraic expressions: get rid of brackets and gather all like terms together.
Give your answer to …	This is an instruction for the accuracy of your final answer. These instructions must be followed or you will lose a mark.
Algebraically	The method you use must involve algebra, e.g. you must solve an equation or simplify an algebraic equation. It is usually stated to avoid trial-and-improvement methods or reading answers from your calculator.
Justify your answer	This is a request for you to indicate clearly your reasoning. Will the examiner know how your answer was obtained?
Show all your working	Marks will be allocated for the individual steps in your working. Steps missed out may lose you marks.

In the Exam

Watch your time and pace yourself carefully. You will find some questions harder than others. Try not to get stuck on one question as you may later run out of time. Rather, return to a difficult question later. Remember that if you have spare time towards the end of your exam, use it to check through your solutions. Mistakes often are discovered in this checking process and can be corrected.

Become familiar with the exam instructions. The practice papers in this book have exam instructions at the front of each exam. Also remember that there is a formuae list to consult. You will find this at the front of your exam paper. However, even though these formulae are given to you, it is important that you learn them so that they are familiar to you. If you are continuing with Mathematics next session it will be assumed that these formulae are known in next year's exam!

Read the question thoroughly before you begin to answer it – make sure you know exactly what the question is asking you to do. If the question is in sections e.g. 15a, 15b, 15c, etc, then it is often the case that answers obtained in the earlier sections will be used in the later sections of that question.

When you have completed your solution read it over again. Is your reasoning clear? Will the examiner understand how you arrived at your answer? If in doubt then fill in more details.

If you change your mind or think that your solution is wrong, don't score it out unless you have another solution to replace it with. Solutions that are not correct can often gain some of the marks available. Do not miss working out. Showing step-by-step working will help you gain maximum marks even if there is a mistake in the working.

Use these resources constructively by reworking questions later that you found difficult or impossible first time round. Remember: success in a Maths exam will only come from actively trying to solve lots of questions and only consulting notes when you are stuck. Reading notes alone is not a good way to revise for your Maths exam. Always be active, always solve problems.

Good luck!

Topic Index

Topic	A Paper 1	A Paper 2	B Paper 1	B Paper 2	C Paper 1	C Paper 2
Unit 1 Expressions and Formulae						
Surds	10		8		11	
Indices	10	1	11		12	1
Algebraic Expressions: Brackets			5			
Algebraic Expressions: Factorising			2		5	
Completing the Square	9			6	9	
Algebraic Fractions	4				5	
Number Sequences						10
Gradient Formula	5		7		1	
Circle Formulae	11					5
Volume of Solids Formulae	1			4		6
Unit 2 Relationships						
Functional Notation			6		3, 10	
Equation of a Sraight Line	5		7		13	
Linear Equations/ Inequations				13	4	11
Simultaneous Equations	7		10		14	
Changing the Subject	11	3		7	6	
Sketching Quadratic Functions	9		12	10		
Equations of Quadratic Functions			12	10		

Quadratic Equations/Discriminant		10, 12	1		10	3
Pythagoras' Theorem & Converse		9		12		8, 12
Similar Shapes				11		12
Trig Graphs	8			8		
Trig Equations		5, 11		8, 9		9
Unit 3 Applications						
Area of a Triangle (Trig)				9		
Sine Rule				3		4
Cosine Rule		8			7	
Bearings Problems (Trig)				3		4
Vectors	3		4	5	8	
Percentages	6	4		1		2
Proportion	6	6				2
Fractions	2		3		2	
Quartiles & Averages		2, 7	9	2		7
Standard Deviation		2		2		7
Scattergraphs & Lines of Best Fit	5		7			

In the answers, there are references to the pages of Leckie & Leckie's Success Guide: National 5 Mathematics. These will help you learn more about any topics you might find difficult.

Practice Paper A

N5

N5 Mathematics

Practice Papers for SQA Exams

Exam A
Paper 1
Non-calculator

Fill in these boxes and read what is printed below.

Full name of centre

Town

Forename(s)

Surname

Attempt ALL of the questions.

You are allowed 1 hour to complete this paper.

Do NOT use a calculator.

Full marks will only be given where your solution contains appropriate working.

Remember to state the units in your answer where appropriate.

Leckie × Leckie
Scotland's leading educational publishers

FORMULAE LIST

The roots of $ax^2 + bx + c = 0$ are $x = \frac{-b \pm \sqrt{(b^2 - 4ac)}}{2a}$

Sine rule: $\frac{a}{\text{Sin } A} = \frac{b}{\text{Sin } B} = \frac{c}{\text{Sin } C}$

Cosine rule: $a^2 = b^2 + c^2 - 2bc \cos A$ or $\cos A = \frac{b^2 + c^2 - a^2}{2bc}$

Area of a triangle: $A = \frac{1}{2}ab \sin C$

Volume of a sphere: $V = \frac{4}{3}\pi r^3$

Volume of a cone: $V = \frac{1}{3}\pi r^2 h$

Volume of a pyramid: $V = \frac{1}{3}Ah$

Standard deviation: $s = \sqrt{\frac{\sum(x - \bar{x})^2}{n - 1}} = \sqrt{\frac{\sum x^2 - (\sum x)^2 / n}{n - 1}}$, where n is the sample size.

MARKS
Do not write in this margin

1. The diagram shows a toy spinner.
The metal cone at the base of the spinner has radius 2 centimetres and height of 6 centimetres as shown.
Calculate the volume of the cone.
Take $\pi = 3{\cdot}14$

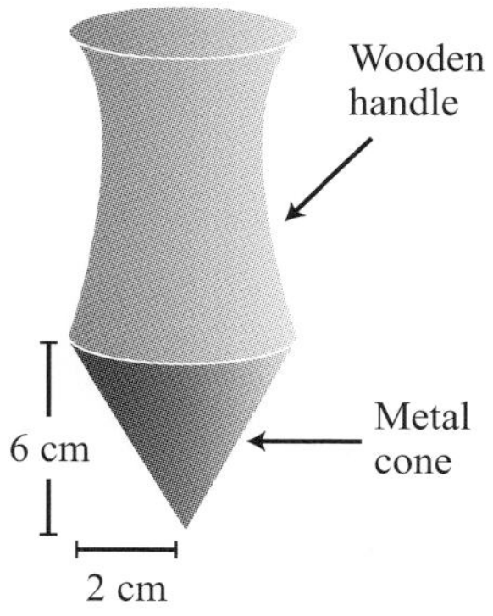

2

2. Evaluate.

$$2\frac{2}{3} - 1\frac{1}{5} \times \frac{1}{3}$$

3

3. Vectors $\boldsymbol{a}$ and $\boldsymbol{b}$ represent two forces acting on a snooker ball.

$$\boldsymbol{a} = \begin{pmatrix} 3 \\ -2 \\ 3 \end{pmatrix} \text{ and } \boldsymbol{b} = \begin{pmatrix} 5 \\ 2 \\ -1 \end{pmatrix}$$

The resultant force is $\boldsymbol{a} + \boldsymbol{b}$. Calculate the magnitude of this force.
Express your answer as a surd in its simplest form. **3**

MARKS Do not write in this margin

4. Express as a single fraction in its simplest form

$$\frac{7}{x}-\frac{3}{x+1}$$ **3**

5. This scattergraph shows how the students in Iain's Physics class scored in their Physics and Maths Exams

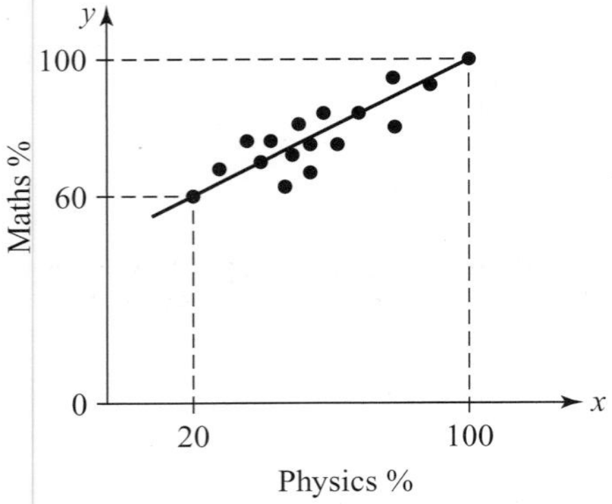

A line of best fit has been drawn.

(a) Find the equation of this line of best fit. **3**

(b) Iain missed his Physics Exam. If he scored 72% in his Maths Exam, use this equation to estimate a score for his Physics Exam. **1**

Total Marks 4

MARKS
Do not write in this margin

6. A retailer is offering a 20% discount on the normal price for all 21-inch 3D televisions.

This is a 21-inch model and is now being sold at a reduced price of £680.
What is the normal price of this television? **3**

MARKS
Do not write in this margin

7. Number Triangles

The rule in a Number Triangle is A = B + C

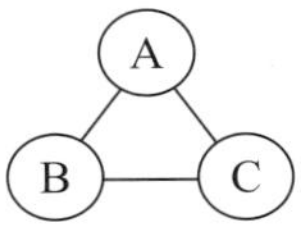

(a) Use this rule to complete this Number Triangle.

Show that $x + 2y = 3$ **2**

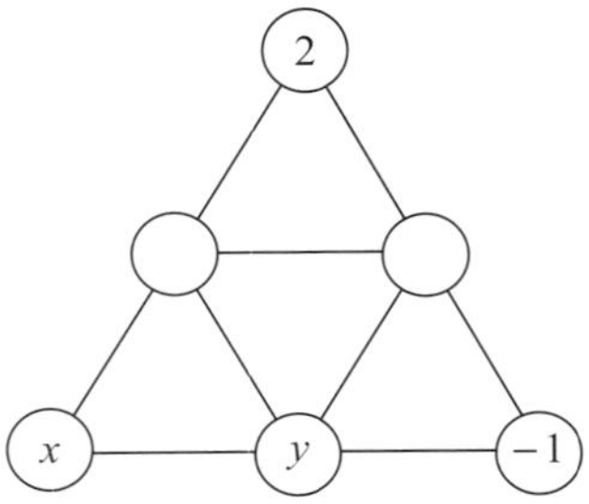

(b) The same Number Triangle rule is used for this Number Triangle.
Write down another equation with x and y. **2**

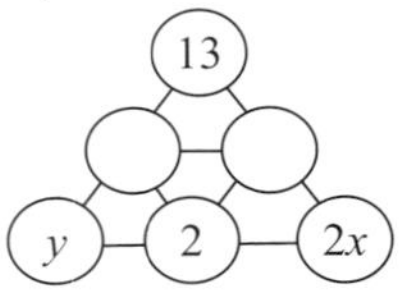

(c) In both these Number Triangles, x and y have the same values.
Find the values of x and y. **3**

Total Marks 7

8. The graph of $y = p \sin qx°$, $0 \leq x \leq 120$ is shown below.

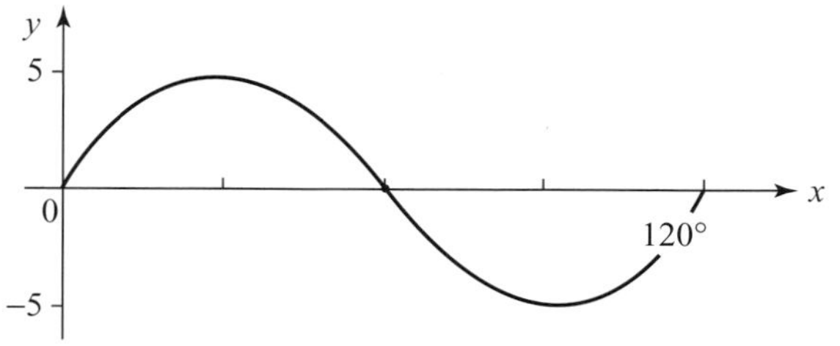

Write down the values of p and q. **2**

9. (a) Write $x^2 + 6x + 16$ in the form $(x + p)^2 + q$ **2**

(b) Sketch the graph of $y = x^2 + 6x + 16$

Show clearly:
- The point of intersection with the y-axis
- The coordinates of the turning point **3**

Total Marks **5**

MARKS
Do not write in this margin

10. (a) Evaluate $8^{\frac{2}{3}} - 8^0$ **2**

(b) Simplify $\sqrt{2} + \sqrt{8}$ **2**

Total Marks 4

11. The circular clock face on Big Ben is set in a square frame as shown in the diagram.

The tip of the minute hand travels a distance of 27 metres every hour.

Show that the perimeter of the square frame is exactly $\frac{108}{\pi}$ metres.

4

[END OF QUESTION PAPER]

N5

N5 Mathematics

Practice Papers for SQA Exams

Exam A
Paper 2

Fill in these boxes and read what is printed below.

Full name of centre

Town

Forename(s)

Surname

Attempt ALL of the questions.

You are allowed 1 hour and 30 minutes to complete this paper.

You may use a calculator.

Full marks will only be given where your solution contains appropriate working.

Remember to state the units in your answer where appropriate.

Leckie × Leckie
Scotland's leading educational publishers

MARKS
Do not write in this margin

1. On 27 August, 2003, Mars was $5{\cdot}6 \times 10^7$ km from Earth, its closest approach for 60 000 years.

How long would it take to drive this distance in a car (if this were possible!) at an average speed of 120 kilometres per hour?

Give your answer in days in scientific notation. **4**

2. A Quality Control Inspector selects a random sample of seven matchboxes produced by Machine A and records the number of matches in each box:

54 45 51 50 48 53 49

(a) For the given data calculate:

(i) the mean **1**

(ii) the standard deviation

Show clearly all your working. **2**

(b) Machine B was also sampled. The data gave a mean of 52 matches and a standard deviation of 1·6 matches. Compare the results for the two machines justifying your comparisons. **2**

Total Marks 5

MARKS Do not write in this margin

3. Change the subject of the formula

$A = \dfrac{m^2 - n}{2}$ to m **3**

4. In 2001, the Russian Space Station, Mir, was destroyed as it burned up in the upper atmosphere.

It had been losing altitude by 6% every month.

At the start of December 2000, its altitude was 340 km.

What was its altitude by the start of March 2001? **3**

5. Solve the equation

$3\sin x^\circ + 2 = 0, \qquad 0 \le x < 360$ **3**

6. The Large Hadron Collider has a circular tunnel with a radius of 4·3 km. In the tunnel protons are accelerated to nearly the speed of light. At this speed they travel 15 km in 51 microseconds (a microsecond is one millionth of a second). In the tunnel there is a detector, Alice, at point A and a detector, LHC 'B', at point B and they form an angle of 96° at the centre of the circle as shown.

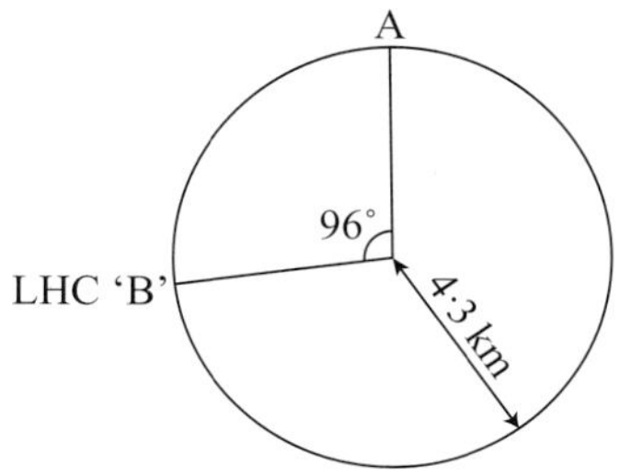

Calculate the time, in microseconds, a proton takes at this speed to travel anti-clockwise from Alice to LHC 'B'. **4**

7. A Biologist was studying the length of time it took bees to return to their hive. These 'out-of-hive' times are shown below in increasing order of time.

5	6	7	7	8	9	12	12	15
18	19	19	19	23	25	34	37	38

(a) Use this data to calculate the interquartile range **2**

(b) For a second hive the interquartile range was found to be 7 minutes.

Make a valid comparison between the hives. **1**

Total Marks 3

8. The diagram shows a gear-slider mechanism.

Rod AB = 10·5 cm

Rod BC = 18 cm

angle ABC = 95°

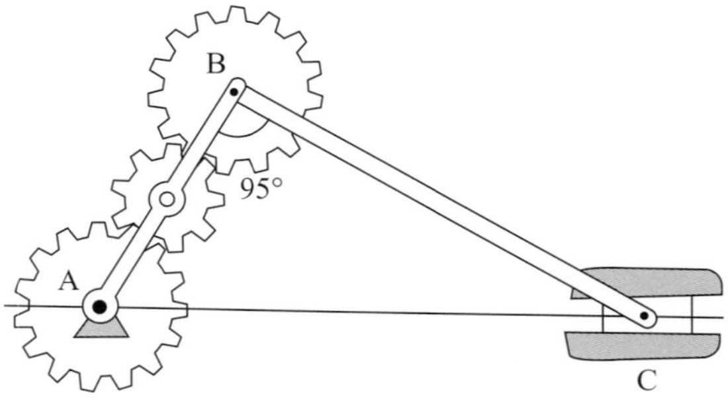

Calculate the length of the slide AC to 1 decimal place. **4**

9. The diagram shows a circular mine shaft with a radius of 5 metres.

A square steel frame is fitted for a lift. A concrete cladding fills the gap between the lift frame and the circular mine shaft.

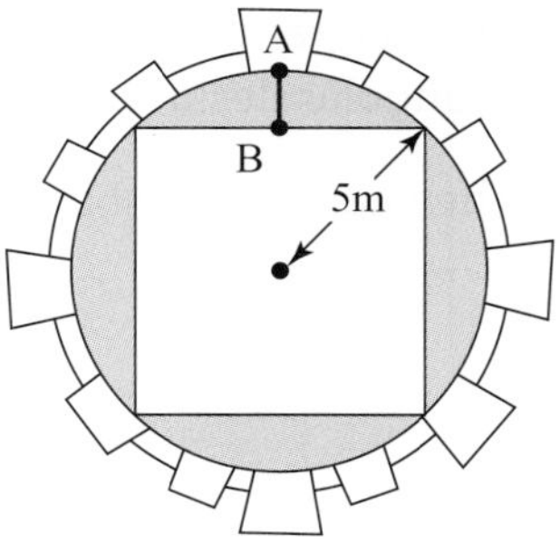

Calculate the greatest width of the concrete cladding (AB in the diagram) giving your answer to 3 significant figures. **5**

10. Find the range of values of k such that the equation $2x^2 + x + k = 0$ has real roots. **4**

MARKS
Do not write in this margin

11. The diagram shows a triangular flag with a shaded triangular design.

The flag is in the shape of an isosceles triangle ACD.

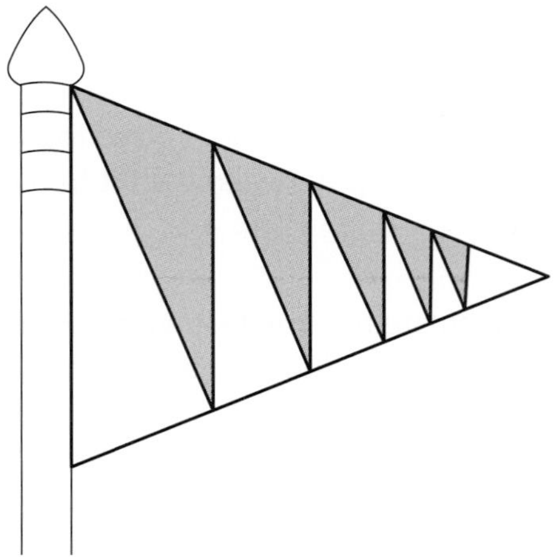

In the design, the largest side of each shaded triangle is at right angles to the lower edge of the flag.

Side AB is 80 cm in length, as shown in the diagram. The tip of the flag forms a 45° angle.

Calculate the length AC of the flag at the flagpole. **5**

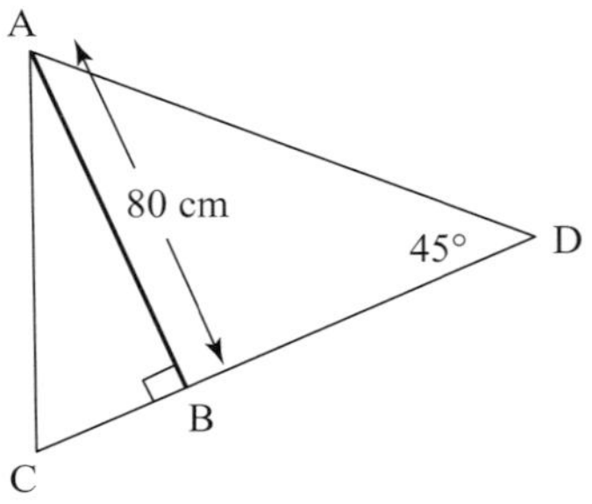

MARKS
Do not write in this margin

12. The diagram shows a swimming pool of width x metres surrounded by a concrete path.

The length of the swimming pool is 2 metres more than its width.

The surrounding concrete path is 1 metre wide along the length and 2 metres wide along the width as shown in the diagram.

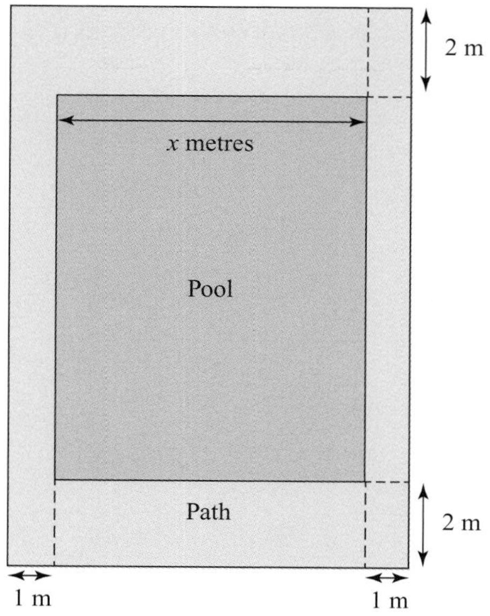

The area of the pool is the same as the area of the concrete path.

(a) Show that $x^2 - 4x - 12 = 0$ **4**

(b) Hence find the dimensions of the pool. **3**

Total Marks **7**

[END OF QUESTION PAPER]

Practice Paper B

FORMULAE LIST

The roots of $ax^2 + bx + c = 0$ are $x = \frac{-b \pm \sqrt{(b^2 - 4ac)}}{2a}$

Sine rule: $\frac{a}{\text{Sin } A} = \frac{b}{\text{Sin } B} = \frac{c}{\text{Sin } C}$

Cosine rule: $a^2 = b^2 + c^2 - 2bc \cos A$ or $\cos A = \frac{b^2 + c^2 - a^2}{2bc}$

Area of a triangle: $A = \frac{1}{2}ab \sin C$

Volume of a sphere: $V = \frac{4}{3}\pi r^3$

Volume of a cone: $V = \frac{1}{3}\pi r^2 h$

Volume of a pyramid: $V = \frac{1}{3}Ah$

Standard deviation: $s = \sqrt{\frac{\sum(x - \bar{x})^2}{n - 1}} = \sqrt{\frac{\sum x^2 - (\sum x)^2 / n}{n - 1}}$, where n is the sample size.

MARKS
Do not write in this margin

1. Ailsa attempts to solve the quadratic equation $2x^2 - 3x - 1 = 0$
She says the discriminant is negative
Do you agree with her? Give a reason for your answer. **2**

2. Factorise

$$2x^2 + 3x - 2$$

2

3. Evaluate

$$\frac{2}{3} \text{ of } \left(1\frac{1}{2} - \frac{1}{3}\right)$$

2

MARKS
Do not write in this margin

4. $\boldsymbol{a} = \begin{pmatrix} -1 \\ -1 \\ 3 \end{pmatrix}$ and $\boldsymbol{b} = \begin{pmatrix} -2 \\ -3 \\ 7 \end{pmatrix}$ are two vectors.

(a) Write down the components of $3\boldsymbol{a}$ **1**

(b) Calculate the components of $3\boldsymbol{a} - 2\boldsymbol{b}$ **1**

(c) Calculate the exact value of $|3\boldsymbol{a} - 2\boldsymbol{b}|$ **1**

Total Marks 3

MARKS Do not write in this margin

5. Simplify

$$x(2x - 1) - x(2 - 3x)$$ **3**

6. $f(x) = \dfrac{6}{x}, x \neq 0$

(a) Evaluate $f(-3)$ **1**

(b) Given that $f(a) = 3$, find a. **2**

Total Marks 3

MARKS
Do not write in this margin

7. The manager of a car insurance company investigates whether the number of telephone operators he employs is related to the number of complaints his company receives on that day.

The results of his investigation are shown in this scattergraph, which also shows a line of best fit.

The two data points A(2, 40) and B(8, 22) lie on this line

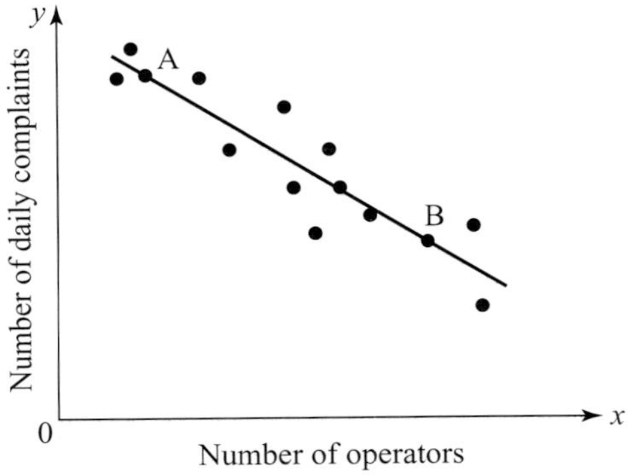

(a) Find the equation of this line of best fit. **3**

(b) If he employs 6 operators, how many complaints should he expect that day? **1**

Total Marks **4**

8. The area of this rectangle is $10\sqrt{2}\,\text{cm}^2$.

It has breadth $\sqrt{10}\,\text{cm}$.

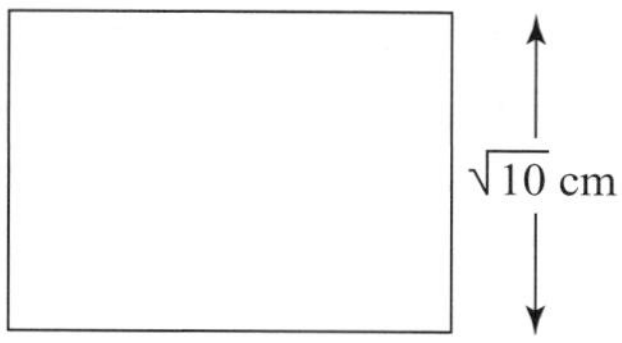

Calculate the length of the rectangle, expressing your answer as a surd in its simplest form. **3**

9. The following data gives the ages of the workers on oil rig Alpha:

21	22	24	23	27	23	23
24	21	22	23	23	21	29
23	19	23	21	24	22	23

(a) For this data find:

(i) The median; **1**

(ii) The lower quartile; **1**

(iii) The upper quartile. **1**

(b) For the workers on oil rig Beta, the interquartile range of their ages is 7 years. Make an appropriate comment on the distribution of the ages of the workers on the two oil rigs. **2**

Total Marks 5

MARKS
Do not write in this margin

10. Protons and neutrons are subatomic particles. Inside each atom there is a nucleus made up from protons and neutrons. These subatomic particles themselves are made up from two types of quarks: 'up' quarks and 'down' quarks. The electrical charge of a proton or neutron is the sum of the electrical charges of its quarks.

(a) A proton is made from two 'up' quarks and one 'down' quark and has a total electrical charge of 1 unit.

Write down an algebraic equation to illustrate this. **1**

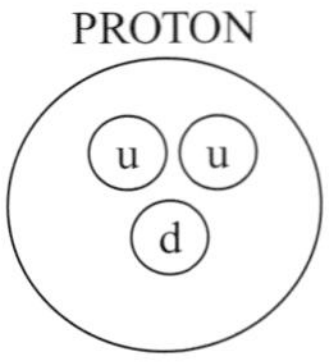

(b) A neutron is made from one 'up' quark and two 'down' quarks and has a total electrical charge of zero.

Write down an algebraic equation to illustrate this. **1**

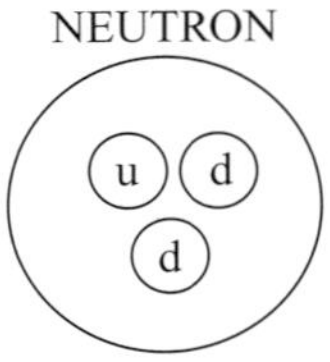

(c) Find the electrical charge of an 'up' quark. Show clearly your reasoning. **2**

Total Marks **4**

MARKS
Do not write in this margin

11. (a) Evaluate $9^{-\frac{1}{2}}$ **2**

(b) Simplify $\sqrt{t} \times t^2$ **2**

Total Marks **4**

MARKS
Do not write in this margin

12. 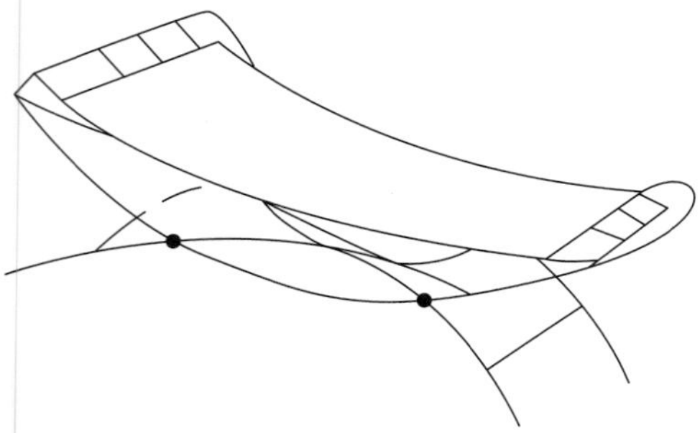

The sketch shows a hammock. Each side of the hammock consists of a metal frame in the shape of two identical parabolas.

Here is a diagram of the frame.

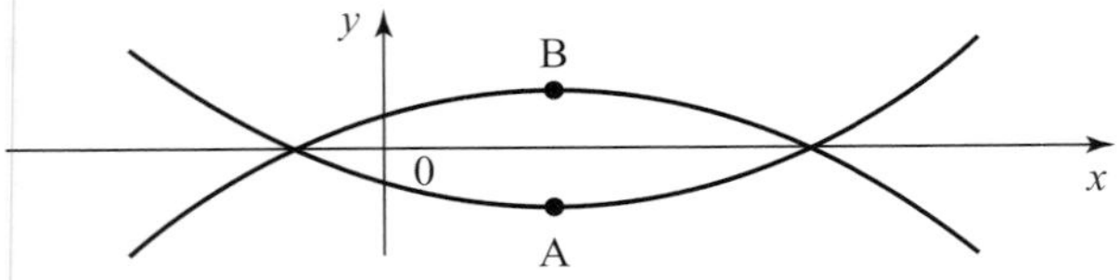

One of these parabolas has equation $y = (x - 2)^2 - 9$. The x-axis is an axis of symmetry for the two parabolas.

(a) State the coordinates of

(i) The turning point A **2**

(ii) The turning point B **1**

(b) Find the equation of the parabola whose equation was not given. **2**

Total Marks 5

[END OF QUESTION PAPER]

MARKS Do not write in this margin

10. The diagram below shows the graph of a quadratic function with the equation

$$y = k(x - p)(x - q) \text{ where } p < q$$

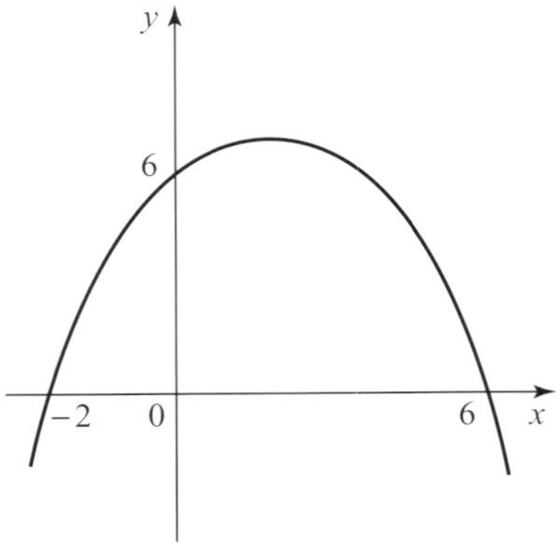

The graph cuts the x-axis at the points $(-2, 0)$ and $(6, 0)$ and cuts the y-axis at the point $(0, 6)$.

(a) Write down the values of p and q. **2**

(b) Calculate the value of k. **2**

(c) Find the coordinates of the maximum turning point of the function. **2**

Total Marks 6

MARKS
Do not write in this margin

11. These two organ pipes are mathematically similar in shape.

The larger pipe is 240 cm in length and the smaller pipe is 180 cm in length.

The volume of the larger pipe is 43 litres.

Calculate the volume of the smaller pipe to the nearest litre. **3**

Practice Exam C

MARKS
Do not write in this margin

8. 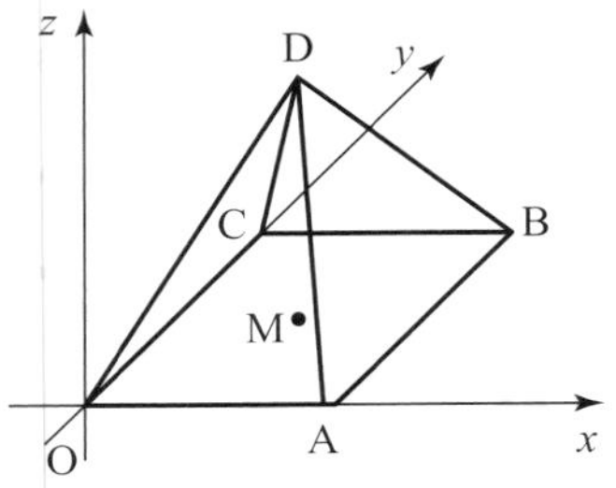

OABCD is a square-based pyramid. Sides OA and OC are placed on the x-axis and y-axis as shown.

O is the origin.

D lies vertically above M, the centre of the square base.

A has coordinates (8, 0, 0)

(a) Find the coordinates of M **1**

(b) If DM = OA find the coordinates of D **1**

(c) $\overrightarrow{AD}$ represents vector $\boldsymbol{w}$ where $\boldsymbol{w} = \begin{pmatrix} -4 \\ 4 \\ 8 \end{pmatrix}$

Calculate $|\boldsymbol{w}|$, the magnitude of $\boldsymbol{w}$.

Express your answer as a surd in its simplest form. **2**

Total Marks 4

MARKS
Do not write in this margin

9. Write $x^2 - 5x + \frac{1}{4}$ in the form $(x - p)^2 + q$ **2**

10. Part of the graphs $y = f(x)$ and $y = g(x)$ are shown where:

$f(x) = 4 + 3x - x^2$

$g(x) = 10 - 2x$

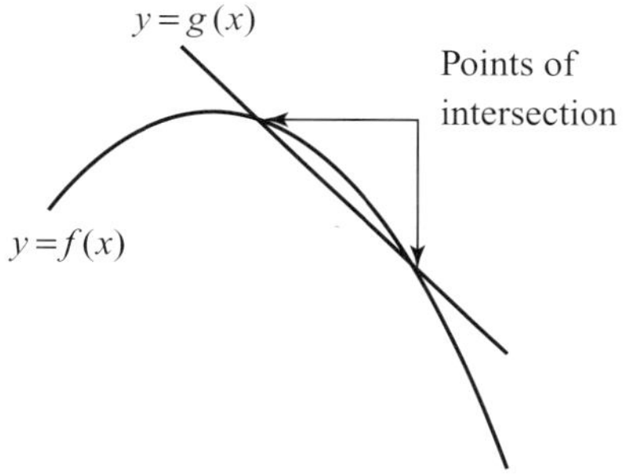

Find the x-coordinates of the two points of intersection by solving the equation $f(x) = g(x)$. **3**

ADDITIONAL SPACE FOR ANSWERS

N5

N5 Mathematics

Practice Papers for SQA Exams

Exam C
Paper 2

Fill in these boxes and read what is printed below.

Full name of centre

Town

Forename(s)

Surname

Attempt ALL of the questions.

You are allowed 1 hour and 30 minutes to complete this paper.

You may use a calculator.

Full marks will only be given where your solution contains appropriate working.

Remember to state the units in your answer where appropriate.

Leckie × Leckie
Scotland's leading educational publishers

MARKS
Do not write in this margin

1. In 2009 scientists at Stanford University broke the record for the smallest writing. They wrote 'SU'. Each letter had width 3×10^{-8} cm.

In letters this size the entire Bible could be written in a line $1{\cdot}05 \times 10^{-1}$ cm long.

Calculate the number of letters in the Bible. Give your answer in scientific notation. **2**

2. Mr Middleton was informed by 'Petcare' that his annual premium for his pet insurance had been increased by $6\frac{1}{2}\%$.

His annual premium was now £87·82.

Calculate his annual premium before this increase. **3**

3. Solve the equation

$3x^2 - x - 5 = 0$

Give your answers correct to 1 decimal place. **4**

MARKS
Do not write in this margin

9. The diagram shows part of the graph $y = \cos x°$

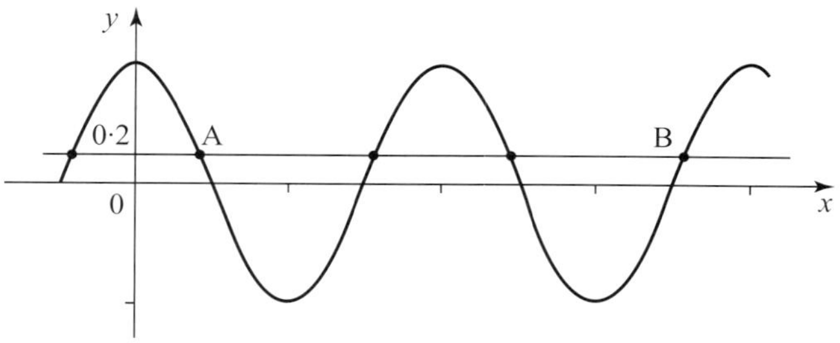

The line $y = 0{\cdot}2$ cuts the graph shown at five points.

Find the x-coordinates of A and B, the 2nd and 5th of these five points. **3**

10. Here is a number pattern:

$2 + 1 = 1 \times 3 - 2 \times 0$

$4 + 1 = 2 \times 4 - 3 \times 1$

$6 + 1 = 3 \times 5 - 4 \times 2$

(a) Write down the 4th line of this pattern. **1**

(b) Write down the *n*th line of this pattern. **2**

(c) Hence show algebraically that this pattern is always true. **3**

Total Marks 6

11. A company is hiring an Excavator machine.

They are considering two Plant Hire companies, 'Earthmove' and 'Trenchers'.

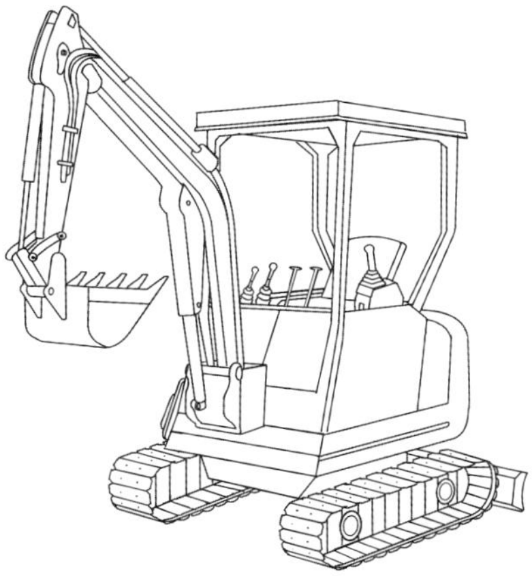

The hire rates for these companies are as follows:

'Earthmove' Delivery/Collection charge £64 Hourly rate: £30	**'Trenchers'** Delivery/Collection charge £28 Hourly rate: £34

(a) Calculate the cost of a 2-hour hire from each of the companies. **1**

(b) For a 20-hour hire which company is cheaper? **1**

(c) For each company find a formula for the cost of a '*n*' hour hire. **2**

(d) 'Trenchers' claim that they are the cheaper of the two companies

Find algebraically the greatest number of hours of hire for this claim to be true. **2**

Total Marks **6**

Answers

Q3.

$$\boldsymbol{a}+\boldsymbol{b}=\begin{pmatrix}3\\-2\\3\end{pmatrix}+\begin{pmatrix}5\\2\\-1\end{pmatrix}=\begin{pmatrix}8\\0\\2\end{pmatrix} \quad ✓$$

$$|\boldsymbol{a}+\boldsymbol{b}|=\sqrt{8^2+0^2+2^2}=\sqrt{64+4}=\sqrt{68} \quad ✓$$

$$\sqrt{68}=\sqrt{4\times17}=\sqrt{4}\times\sqrt{17}=2\sqrt{17} \quad ✓$$

3 marks

Addition
- Correct components will gain 1st mark

Magnitude
- You have to know that the magnitude of a vector with components $\begin{pmatrix}p\\q\\r\end{pmatrix}$ is given by $\sqrt{p^2+q^2+r^2}$

SUCCESS GUIDE: page 116

Simplification
- Remember to look for factors that are square numbers: 4, 9, 16, …
- $\sqrt{17}$ cannot be simplified further since 17 is a prime number and has no factors other than 1 and 17.

SUCCESS GUIDE: page 9

Q4. $\frac{7}{x}-\frac{3}{x+1}$

$$=\frac{7(x+1)}{x(x+1)}-\frac{3x}{x(x+1)} \quad ✓$$

$$=\frac{7(x+1)-3x}{x(x+1)}$$

$$=\frac{7x+7-3x}{x(x+1)}$$

$$=\frac{4x+7}{x(x+1)} \quad ✓$$

2 marks

Subtraction of algebraic fractions
- A common denominator is needed: multiply x and $x+1$ together
- Always multiply the 'top' and the 'bottom' by the same expression: $\frac{7\times(x+1)}{x\times(x+1)}$ and $\frac{3\quad\times x}{x+1\times x}$

Simplify
- The numerators are subtracted and then simplified

SUCCESS GUIDE: page 34

Q5. (a) Two points on the line are A(20, 60) and B(100, 100)

so $m_{AB}=\frac{100-60}{100-20}=\frac{40}{80}=\frac{1}{2}$ ✓

now use $y-b=m(x-a)$

with $m=\frac{1}{2}$ and (a, b) as (20, 60)

so the equation is ✓

$$y-60=\frac{1}{2}(x-20)$$

$$\Rightarrow y-60=\frac{1}{2}x-10$$

$$\Rightarrow y=\frac{1}{2}x+50 \quad ✓$$

3 marks

Gradient
- Use the formula $m=\frac{y_2-y_1}{x_2-x_1}$

SUCCESS GUIDE: pages 36, 37

Substitution
- Alternatively $y=mx+c$ may be used giving $60=\frac{1}{2}\times20+c$ (leading to $c=50$)

Equation
- Alternative forms are: $2y=x+100$ or $2y-x=100$

SUCCESS GUIDE: pages 56–58

Q5. (b) The Maths scores are the y values so let $y = 72$

Now solve $72 = \frac{1}{2}x + 50$

$\Rightarrow \frac{1}{2}x = 22 \Rightarrow x = 44$ ✓

His estimate is 44%

1 mark

Solve equation
- Be careful to read the descriptions on the axes. A common mistake here is to substitute $x = 72$.

SUCCESS GUIDE: pages 57, 58

Q6. Normal price (100%) has been reduced by 20% to 80%:

$80\% \longleftrightarrow £680$

$1\% \longleftrightarrow \frac{680}{80}$ ✓

$100\% \longleftrightarrow \frac{680}{80} \times 100$ ✓

$= £850$ ✓

3 marks

Percentages
- The final price is given *after* a reduction so a 'proportion' method is required.
- Knowing 80%, divide by 80 to find 1% and multiply by 100 to find 100%.

SUCCESS GUIDE: pages 119, 120

Q7. (a) This gives: ✓

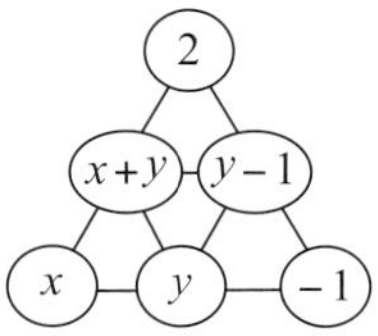

$x + y + y - 1 = 2$

$x + 2y - 1 = 2$

$x + 2y = 2 + 1$

$x + 2y = 3$ ✓

2 marks

Understanding the pattern
- The example A = B + C tells you that the top circle is the sum of the two below it: (x) + (y) and (y) + (-1)

Set up an equation
- Following the pattern for the top triangle leads to an equation: ($x + y$) + ($y - 1$) = (2)

Q7. (b) This gives: ✓

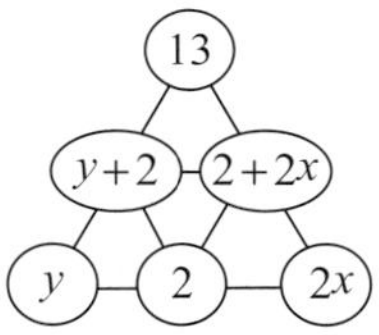

$y + 2 + 2 + 2x = 13$

$y + 4 + 2x = 13$

$x + 2x = 13 - 4$ ✓

$y + 2x = 9$

2 marks

Set up an equation
- Following the pattern for the top triangle leads to an equation: ($x + y$) + ($y - 1$) = (2)

Simplify the equation
- Letters on the left, numbers on the right

Q7. (c) Solve simultaneous equations

$$\left.\begin{array}{l} x + 2y = 3 \\ 2x + y = 9 \end{array}\right\} \begin{array}{l} \times 2 \longrightarrow 2x + 4y = 6 \\ \longrightarrow \underline{2x + y \;\; = 9} \end{array}$$ ✓

subtract $3y = -3$

so $y = -1$ ✓

Put $y = -1$ into $2x + y = 9$

so $2x - 1 = 9 \Rightarrow 2x = 10$

$\Rightarrow x = 5$ ✓

3 marks

Simultaneous equations
- You will recognise this from having two equations and two unknowns (x and y)
- Line up the equations x's under x's and y's under y's with numbers on the right.

Calculation of values
- Aim for one equation with one unknown
- Substitute the first calculated value back into either of the two equations
- Check using the other equation. Use $y = -1$ and $x = 5$ in $x + 2y$ to get $5 + 2x(-1) = 3$

SUCCESS GUIDE: page 61

Q8. The amplitude is 5 so
$p = 5$ ✓

From 0° to 360° there would be 3 cycles so $q = 3$ ✓

2 marks

Amplitude
- $y = \sin x°$ has a max or min value of 1 or −1 so $y = 5\sin x°$ has a max or min value of 5 or −5, true for this graph.

Period
- $y = \sin x°$ has period 360° – 1 cycle every 360°. So $y = \sin 3x°$ has period 120° – 3 cycles every 360°; true for this graph.

SUCCESS GUIDE: page 94

Q9. (a) Compare $x^2 + 6x + 16$
with $(x + 3)^2 = x^2 + 6x + 9$ ✓
so $x^2 + 6x + 16$
$= (x + 3)^2 - 9 + 16$ ✓
$= (x + 3)^2 + 7$

2 marks

Value of *p*
- The coefficient of x is halved: $\frac{1}{2}$ of $6 = 3$

Value of *q*
- When $(x + 3)^2$ is multiplied out, the constant term 9 is subtracted and the required constant 16 added.

SUCCESS GUIDE: page 28

Q9. (b)

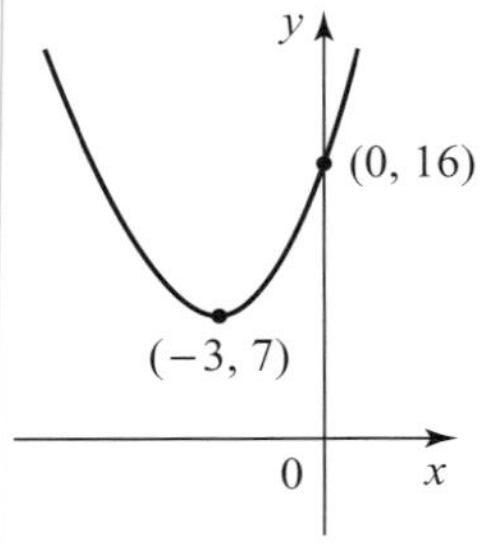

✓ ✓ ✓

3 marks

Parabolic shape
- $y = x^2$ is shifted left and moved up

***y*-intercept**
- set $x = 0$ in $y = x^2 + 6x + 16$ to get $y = 16$

Turning point
- The shift to $y = x^2$ is 3 left and 7 up.

SUCCESS GUIDE: page 71

Q10. (a) $8^{\frac{2}{3}} - 8^0$ ✓

$= (\sqrt[3]{8})^2 - 1$
$= 2^2 - 1 = 4 - 1 = 3$ ✓

2 marks

Zero index
- $a^0 = 1$ Any non-zero number raised to the power zero gives 1.

Fractional indices
- $a^{\frac{m \leftarrow \text{power}}{n \leftarrow \text{root}}}$ so $a^{\frac{m}{n}} = (\sqrt[n]{a})^m$ and in this case: $8^{\frac{2 \leftarrow \text{squared (power 2)}}{3 \leftarrow \text{cube root}}}$

SUCCESS GUIDE: page 18

Q10. (b)
$\sqrt{2} + \sqrt{8}$
$= \sqrt{2} + \sqrt{4 \times 2}$ ✓
$= \sqrt{2} + \sqrt{4} \times \sqrt{2}$
$= \sqrt{2} + 2\sqrt{2} = 3\sqrt{2}$ ✓

2 marks

Simplifying surds
- Always attempt to get the smallest number under the square root by factorising out squares: 4,9,16,...

Adding surds
- Compare $= \sqrt{2} + 2\sqrt{2} = 3\sqrt{2}$ with $x + 2x = 3x$

SUCCESS GUIDE: pages 9, 11

Q11. In 1 hour the tip of the minute hand travels the whole circumference of the clock face ✓

so $\pi D = 27$ ✓

$\Rightarrow D = \dfrac{27}{\pi}$ ✓

The diameter, D, of the clock face is the same length as the side of the square frame

so perimeter $= 4 \times D$

$= 4 \times \dfrac{27}{\pi}$

$= \dfrac{108}{\pi}$ metres ✓

4 marks

Strategy

- Evidence that you knew to use the Circumference formula to set up an equation will gain you the strategy mark.

Set up equation

- $C = 2\pi r$ or $C = \pi D$: the second version is the best one for this question.

SUCCESS GUIDE: page 38

Change the subject

- Divide both sides by π

Calculation

- 4 sides make up the perimeter, so $\times$ 4
- The working $4 \times \dfrac{27}{\pi}$ is essential especially since the answer $\dfrac{108}{\pi}$ is given.

Exam A: Paper 2

Q1. Use $T = \dfrac{D}{S}$

so $T = \dfrac{5{\cdot}6 \times 10^7}{120}$ ✓

$= 466666{\cdot}66\ldots$ hours ✓

$= \dfrac{466666{\cdot}66\ldots}{24}$ days

$= 19444{\cdot}44\ldots$ ✓

$= 1{\cdot}9 \times 10^4$ days (to 2 sig figs)

4 marks

***D,S,T* formula**

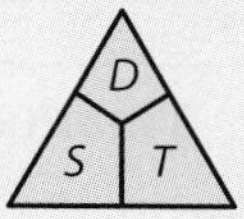

- You are asked to find the Time taken. Covering T in the triangle gives $\dfrac{D}{S}$, the Distance divided by the Speed.

Calculator

- $5{\cdot}6 \times 10^7$ is entered into your calculator like this:

[5] [·] [6] [×10^x] [7] [=]

Conversion

- Change hours to days: divide by 24.

Scientific notation

- Convert your answer to scientific notation:

$a \times 10^n$

a number between 1 and 10, in this case 1·9

The decimal point in 1·9 moves 4 places to the right so $n = 4$

SUCCESS GUIDE: pages 18, 19

Q2. (a)(i)

$$\text{Mean} = \frac{54+45+51+50+48+53+49}{7}$$

$$= \frac{350}{7} = 50 \quad ✓$$

1 mark

Mean

- $\text{Mean} = \dfrac{\text{sum of numbers}}{\text{number of numbers}} = \dfrac{\sum x}{n} = \bar{x}$

Q2. (a)(ii)

x	$x-\bar{x}$	$(x-\bar{x})^2$
54	4	16
45	−5	25
51	1	1
50	0	0
48	−2	4
53	3	9
49	−1	1

✓

$\sum(x-\bar{x})^2 = 56$ ✓

$$s = \sqrt{\frac{\sum(x-\bar{x})^2}{n-1}} = \sqrt{\frac{56}{6}} \doteqdot 3{\cdot}1 \text{ (to 1 dec.pl.)}$$

2 marks

Squared deviations

- This mark is gained for the correct last column in the table: the squared values of the deviations from the mean.

Substitution & calculation

- The formula $s = \sqrt{\frac{\sum(x-\bar{x})^2}{n-1}}$ is given to you in the exam on your formulae page.
- This mark is for correctly substituting $\sum(x-\bar{x})^2 = 56$ and $n - 1 = 6$ into the formula and completing the calculation.
- Using the other formula gives $\sum x = 350$, $\sum x^2 = 17556$ and $n = 7$, resulting in $s \doteqdot 3{\cdot}1$.
- If you know how to use the STAT mode on your calculator to check $s = 3{\cdot}1$ then you should do this. Remember just writing the answer down from the calculator with no working will not gain you the marks.

SUCCESS GUIDE: pages 130, 131

Q2. (b)

On average there are more matches in Machine B's boxes (mean = 52 is greater than mean = 50 for machine A). ✓

There is much more variation in the number of matches in Machine A's boxes than those from Machine B ✓
($s = 3{\cdot}1$ for Machine A is greater than $s = 1{\cdot}6$ for Machine B).

2 marks

1st statement

- This concerns the 'average' contents. Back up your statement with the statistics (52 and 50).

2nd statement

- This concerns the distribution about the mean. A larger standard deviation means more variation about the mean.

Q3. $A = \dfrac{m^2 - n}{2}$

$(\times 2)\ (\times 2)$ ✓

$\Rightarrow 2A = m^2 - n$

$(+n)\quad (+n)$ ✓

$\Rightarrow 2A + n = m^2$

So $m^2 = 2A + n$

$\Rightarrow m = \sqrt{2A + n}$ ✓

3 marks

1st step
- Get rid of the fraction: multiply both sides of the equation by 2.

2nd step
- Try to isolate the expression involving the subject m: add n to both sides

Last step
- The 'inverse' of squaring is 'square rooting' so take the square root of both sides.

SUCCESS GUIDE: pages 64, 65

Q4. Reduction factor is 0·94 and is applied three times. ✓

So Final altitude
= 340 × 0·94 × 0·94 × 0·94 ✓
= 282·39... ≑ 282 km
(to nearest km) ✓

3 marks

Reduction factor
- Reducing by 6% is equivalent to calculating 94%, i.e. multiplying by 0·94.

SUCCESS GUIDE: page 119

Time calculation
- Shorter version is $340 \times 0{\cdot}94^3$

Rounding
- Unless accuracy is stated in the question any correct rounding is acceptable.

Q5. $3 \sin x° + 2 = 0$

$\Rightarrow 3 \sin x° = -2 \Rightarrow \sin x° = -\dfrac{2}{3}$ ✓

$x°$ is in the 3rd or 4th quadrants the 1st quadrant angle is 41·8°

So $x = 180 + 41{\cdot}8$
$= 221{\cdot}8$ ✓
or
$x = 360 - 41{\cdot}8$
$= 318{\cdot}2$ ✓

3 marks

Rearrangement
- The 1st step is to find the value of $\sin x°$

1st value
- $\sin x°$ is negative: use

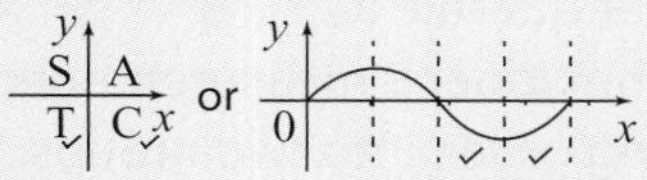

to determine the quadrants (3rd and 4th)
- 1st quadrant angle is obtained from $\sin^{-1}\left(\dfrac{2}{3}\right)$. Do not use the negative with $\boxed{\sin^{-1}}$
- 3rd quadrant angle is always 180° + (1st quad angle)

2nd value
- 4th quadrant angle is always 360° − (1st quad angle)
- Use this diagram to remember how to get the different quadrant angles (θ is the 1st quadrant angle)

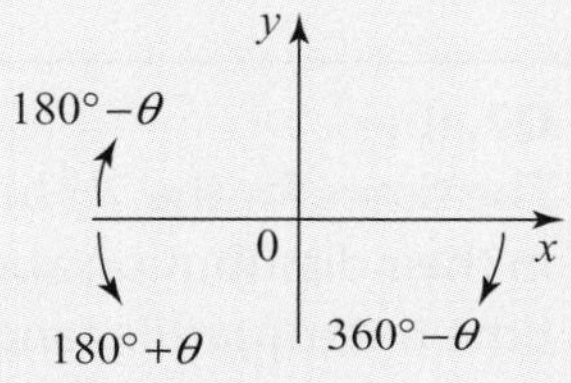

SUCCESS GUIDE: pages 96, 97

Q6. Length of arc AB

$= \frac{96}{360} \times$ circumference ✓

$= \frac{96}{360} \times \pi \times 8 \cdot 6 = 7 \cdot 204...\text{km}$ ✓

So 15 km $\longleftrightarrow$ 51 microsec

$\Rightarrow 1 \text{ km} \longleftrightarrow \frac{51}{15}$ microsec

$\Rightarrow 7{\cdot}204\ldots \longleftrightarrow \frac{51}{15} \times 7{\cdot}204$ microsec

$= 24{\cdot}496$ ✓

The proton takes 24·5 microsec to travel from A to B in the tunnel. ✓

4 marks

Strategy
- The angle at the centre of the circle is in direct proportion to the length of the arc. So we have
- 96° $\longleftrightarrow$ arc AB
- 360° $\longleftrightarrow$ whole circumference
- This leads to $\frac{96}{360}$ as the required fraction of the circumference that gives the length of arc AB.
- This mark is for the appearance of $\frac{96}{360}$ or equivalent.

Arc length
- $C = \pi D$ where $D = 2 \times$ radius
  $= 2 \times 4{\cdot}3$ km

Method
- A proportion problem: Divide by 15 then multiply by 7·204…. Evidence that you knew this will gain you this mark.

Solution
- Correct calculation gains the final mark.

SUCCESS GUIDE: page 38

Q7. (a) There are 18 measurements. Divide them, in increasing order, into two groups of 9:

(5...........15)(18...........38)

Divide the lower 9 measurements, in increasing order, into two groups:

(5 6 7 7) 8 (9 12 12 15)
↑

The lower quartile (Q_1) = 8

Divide the upper 9 measurements, in increasing order, into two groups:

(18 19 19 19) 23 (25 34 37 38)
↑ ✓

The upper quartile (Q_3) = 23

interquartile range

$= Q_3 - Q_1 = 23 - 8 = 15.$ ✓

2 marks

Lower quartile & Upper quartile
- Another way of thinking about the lower quartile is that it is the median of the lower half of the measurements.
- The upper quartile can be thought of as the median of the upper half of the measurements.
- Remember that the symbol Q_2 is reserved for the median of the whole set of data.

Interquartile range
- The formula Int. Range $= Q_3 - Q_1$ is not given to you in the Exam So you need to remember it.

SUCCESS GUIDE: pages 127–129

Q7. (b)

The times for the 2nd hive are less varied in their distribution about the median time (interquartile range is 7) than for the 1st hive (interquartile range is 15 which is greater than 7). ✓

1 mark

Statement
- In your statement you must compare the two interquartile ranges 7 and 15 otherwise you will not gain this mark.

SUCCESS GUIDE: page 128

Q8. Use the cosine rule

$b^2 = a^2 + c^2 - 2ac \cos B$ ✓

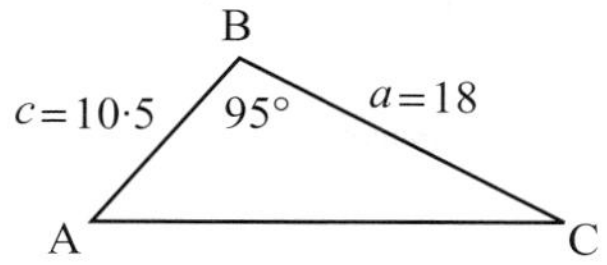

✓

$= 18^2 + 10{\cdot}5^2 - 2 \times 18 \times 10{\cdot}5 \times \cos 95°$

$= 467{\cdot}19...$ ✓

so $b = \sqrt{467{\cdot}19...} = 21{\cdot}61...$

so AC ≑ 21·6 cm (to 1 dec. place) ✓

4 marks

Strategy
- Two sides and the included angle are given so the cosine rule is used.

Substitution
- Correct substitution of the values into the formula will gain you a mark.

Calculation
- In most calculators just key in the complete calculation as written in the solution.

Rounding
- 1 mark is allocated for correct rounding.

SUCCESS GUIDE: page 108

Q9.

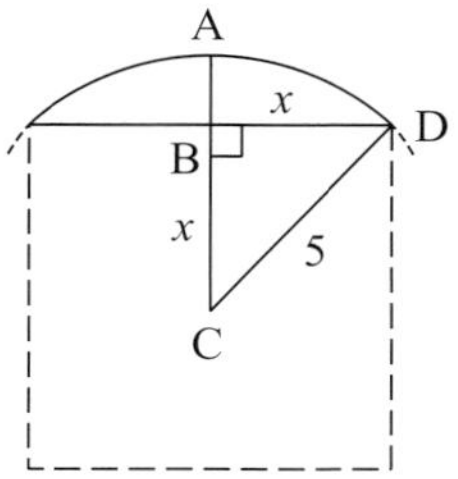

Triangle BCD (see diagram) is isosceles with BC = BD = x ✓

✓

By Pythagoras' Theorem: ✓

$x^2 + x^2 = 5^2$

$\Rightarrow 2x^2 = 25$

$\Rightarrow x^2 = \frac{25}{2} = 12{\cdot}5$ ✓

So $x = \sqrt{12{\cdot}5}$. AC is a radius

so AC = 5 ✓

AB = AC − BC = $5 - \sqrt{12{\cdot}5}$

= 5 − 3·53... = 1·464...

So AB ≑ 1·46 cm (to 3 sig figs)

5 marks

Strategy
- Constructing the right-angled triangle and using Pythagoras' Theorem is the crucial strategy.

Pythagoras' Theorem
- Naming BC and BD with letter x helps.

Strategy
- Realisation that AC is a radius, length 5 m, is the 2nd crucial strategy step.

Calculation
- Never round answers until the final answer.

Rounding
- Rounding to 3 significant figures is essential.

SUCCESS GUIDE: pages 76, 84, 85

Q10. Compare $2x^2 + x + k = 0$

with $ax^2 + bx + c = 0$

This gives $a = 2$, $b = 1$ and $c = k$ ✓

so Discriminant $= b^2 - 4ac$

$= 1^2 - 4 \times 2 \times k$ ✓

$= 1 - 8k$

For real roots Discriminant ≥ 0

so $1 - 8k \geq 0$ ✓

$\Rightarrow 1 \geq 8k \Rightarrow \frac{1}{8} \geq k$

So the range is $k \leq \frac{1}{8}$ ✓

4 marks

Values
- This is for obtaining the values of the coefficients a, b and c.

Discriminant
- This mark is for knowing to use the discriminant $b^2 - 4ac$.

Inequation
- No real roots if discriminant < 0, real roots in all other cases.

Solution
- Add $8k$ to both sides of the inequation then divide both sides by 8.
- Notice that $a \leq b$ is the same as $b \geq a$.

SUCCESS GUIDE: page 74

Q11. The large triangle is isosceles

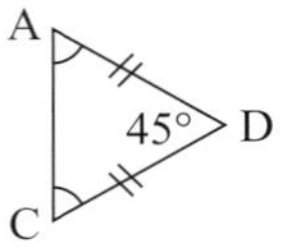

So angle A = angle C ✓

angle C $= \frac{1}{2}(180° - 45°)$

$= \frac{1}{2} \times 135° = 67{\cdot}5°$ ✓

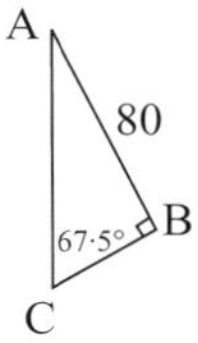

Now use 'SOHCAHTOA' in triangle ABC ✓

$\sin 67{\cdot}5° = \frac{80}{AC}$ ✓

$\Rightarrow AC \times \sin 67{\cdot}5° = 80$

$\Rightarrow AC = \frac{80}{\sin 67{\cdot}5°} = 86{\cdot}59\ldots$

$\doteqdot$ 86·6 cm ✓
(to 3 sig. figs)

5 marks

Strategy
- Evidence that you knew to use 'SOHCAHTOA' in triangle ABC by finding angle C would gain you the strategy mark

Angle sum
- Angle A and angle C between them make up 135° of the 180° angle sum in the large triangle.

Correct ratio
- Working from angle C: the opposite AB is known and the hypotenuse AC is required, so use 'SOH', i.e. sin.

Correct equation
- $\sin 67{\cdot}5 = \frac{80}{AC}$ or its equivalent will gain you a further mark.

Calculation
- Multiply both sides of the equation by AC and then divide both sides by sin 67·5°.

SUCCESS GUIDE: pages 89, 90

Q12. (a) Use Area = length × breadth

Pool: Outside of Path:

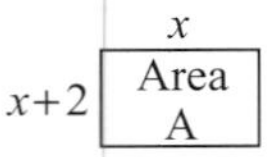

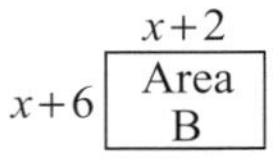

✓

Area $= x(x+2)$ Area $= (x+2)(x+6)$
$= x^2 + 2$ $= x^2 + 8x + 12$ ✓

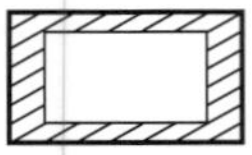

Area of Path = Area B − Area A

$= x^2 + 8x + 12 - (x^2 + 2x)$
$= x^2 + 8x + 12 - x^2 - 2x$ ✓
$= 6x + 12$

Area of Pool = Area of Path

$\Rightarrow x^2 + 2x = 6x + 12$
$\Rightarrow x^2 + 2x - 6x - 12 = 0$ ✓
$\Rightarrow x^2 - 4x - 12 = 0$

4 marks

Starting
- You will gain a mark for either of the two area expressions appearing.

Strategy
- Any correct method for finding an expression for the area of path and then setting up an equation will guarantee the strategy mark.

Simplification
- Simplifying expressions as you proceed through your solution makes the subsequent working easier.

Equation rearrangement
- Since the answer is given ($x^2 - 4x - 12 = 0$) it is important that your working is clear.

Q12. (b) $x^2 - 4x - 12 = 0$

$\Rightarrow (x-6)(x+2) = 0$ ✓
$\Rightarrow x = 6$ or $x = -2$

$x = -2$ is not sensible for a length
So $x = 6$ is the only solution. ✓
Pool dimensions:
6 m × 8 m ✓

3 marks

Solving a Quadratic
- Each factor could be zero: $x - 6 = 0$ or $x + 2 = 0$ giving the two solutions $x = 6$ and $x = -2$.

Invalid Solution
- You must state clearly that $x = -2$ makes no sense. There is a mark for this.

Answer
- The dimensions of the pool are x metres by $(x + 2)$ metres with $x = 6$.

SUCCESS GUIDE: pages 67, 68

Answers to Exam B

Exam B: Paper 1

Q1. Compare $2x^2 - 3x - 1 = 0$

with $ax^2 + bx + c = 0$

This gives $a = 2, b = -3$ and $c = -1$

Discriminant $= b^2 - 4ac$

$= (-3)^2 - 4 \times 2 \times (-1)$ ✓

$= 9 + 8 = 17$

This is positive so she is wrong. ✓

2 marks

Substitution

- For correct substitution into $b^2 - 4ac$

Calculation & statement

- Remember negative × negative gives a positive result: $(-3)^2 = -3 \times (-3) = 9$ and $-4 \times 2 \times (-1) = 8$
- You will lose this mark if you fail to mention that 17 is positive (or non-negative).

Q2. $2x^2 + 3x - 2$

$= (2x - 1)(x + 2)$ ✓✓

2 marks

Factorisation

- You should always check your answer by multiplying out using 'FOIL'.

SUCCESS GUIDE: pages 26, 27

Q3. $\frac{2}{3}$ of $\left(1\frac{1}{2} - \frac{1}{3}\right)$

$= \frac{2}{3}$ of $\left(\frac{3}{2} - \frac{1}{3}\right)$

$= \frac{2}{3}$ of $\left(\frac{9}{6} - \frac{2}{6}\right) = \frac{2}{3} \times \frac{9-2}{6}$ ✓

$= \frac{2}{3} \times \frac{7}{6} = \frac{14}{18} = \frac{7}{9}$ ✓

2 marks

Subtraction of fractions

- The lowest common denominator is 6 as this is the smallest number 2 and 3 divide into exactly.

Multiplication of fractions

- "Multiply across": $\frac{a}{b} \times \frac{c}{d} = \frac{a \times c}{b \times d}$
- Cancel if possible: in this case divide top and bottom of the fraction by 2 to get $\frac{7}{9}$.

SUCCESS GUIDE: pages 122, 123

Q4. (a)

$$3\mathbf{a} = 3\begin{pmatrix} -1 \\ -1 \\ 3 \end{pmatrix} = \begin{pmatrix} 3\times(-1) \\ 3\times(-1) \\ 3\times3 \end{pmatrix} = \begin{pmatrix} -3 \\ -3 \\ 9 \end{pmatrix}$$ ✓

1 mark

Components

- Each component is trebled

Q4. (b)

$$3\mathbf{a} - 2\mathbf{b} = \begin{pmatrix} -3 \\ -3 \\ 9 \end{pmatrix} - 2\begin{pmatrix} -2 \\ -3 \\ 7 \end{pmatrix} = \begin{pmatrix} -3+4 \\ -3+6 \\ 9-14 \end{pmatrix} = \begin{pmatrix} 1 \\ 3 \\ -5 \end{pmatrix}$$ ✓

1 mark

Components

- Take care with negative signs

$-2\begin{pmatrix} -2 \\ -3 \\ 7 \end{pmatrix}$ ←positive $-2\times(-2)$; ←positive $-2\times(-3)$; ←negative -2×7

Q9. (b) For oil rig Alpha:
Interquartile range
$= 23{\cdot}5 - 21{\cdot}5 = 2$ ✓
The age distribution is more spread out (about the mean age) on oil rig Beta compared to that on oil rig Alpha since the interquartile range of 7 for Beta is greater than the 2 for Alpha. ✓

2 marks

Interquartile range
- You need to calculate the statistic: Interquartile range $= Q_3 - Q_1$ for oil rig Alpha and then compare it to the same statistic for oil rig Beta.

Comparison
- It is important you back up any comparison statement with the relevant statistics. In this case the interquartile range: the greater the range the more spread out the data is around the median.

SUCCESS GUIDE: pages 128, 129

Q10. (a) Let u be the charge on an 'up' quark and let d be the charge on a 'down' quark:
$2u + d = 1$ ✓

1 mark

Setting up an equation
- It is important to be clear what the letters you use are representing. Here it is the value of the electric charge.

Q10. (b) $u + 2d = 0$ ✓

1 mark

Setting up an equation
- 'Algebraic' mentioned in a question means that algebra should be used, e.g. equation solving, letter substitution, etc.

Q10. (c) Solve simultaneous equations ✓

$$\left.\begin{aligned} 2u+d&=1 \\ u+2d&=0 \end{aligned}\right\} \begin{aligned} &\times 2 \rightarrow & 4u+2d&=2 \\ &\longrightarrow & u+2d&=0 \end{aligned}$$ ✓

subtract $3u = 2$

$$\Rightarrow u = \frac{2}{3}$$ ✓

2 marks

Strategy
- Two equations with two unknowns (letters) indicates simultaneous equations need to be solved to find the values of the unknowns.

Method
- Line up the equations: in this case 'u's are lined up and 'd's are lined up.
- The aim is always to eliminate one of the unknowns, in this case d.

Calculation
- Divide both sides by 3.

SUCCESS GUIDE: page 61

Q11. (a) $9^{-\frac{1}{2}} = \dfrac{1}{9^{\frac{1}{2}}}$ ✓

$= \dfrac{1}{\sqrt{9}} = \dfrac{1}{3}$ ✓

2 marks

Negative indices
- $a^{-n} = \dfrac{1}{a^n}$ is the general Index Law used

Fractional indices
- $a^{\frac{m}{n}} = \left(\sqrt[n]{a}\right)^m$ is the general Index Law used.
This may help:
$a^{\frac{m}{n}}$ (m ← power, n ← root) so $a^{\frac{1}{2}}$ (1 ← power 1, 2 ← square root)

So $9^{\frac{1}{2}}$ means the square root of 9 to the power 1.

SUCCESS GUIDE: pages 17, 18

Q11. (b) $\sqrt{t} \times t^2$

$= t^{\frac{1}{2}} \times t^2$ ✓

$= t^{\frac{1}{2}+2} = t^{\frac{5}{2}}$ ✓

2 marks

Simplifying indices

- $a^m \times a^n = a^{m+n}$ is the Index law used
- Note $t^{2\frac{1}{2}}$ is not acceptable.

SUCCESS GUIDE: page 17

Q12. (a) (i) ✓

A = (2, − 9) ✓

2 marks

***x*-coordinate**

- The least value of $(x - 2)^2 - 9$ is $0^2 - 9$ and this occurs when $x = 2$. Remember something squared must be zero or positive, never a negative value.

***y*-coordinate**

- when $x = 2$, $y = (2 - 2)^2 - 9 = 0^2 - 9 = -9$

SUCCESS GUIDE: page 71

Q12. (a) (ii)

B = (2, 9) ✓

1 mark

Turning point

- By symmetry point B is the reflection of point A in the x-axis so the x-coordinate will be the same (2) and the y-coordinate positive 9.

Q12. (b) The other parabola is a reflection of $y = (x - 2)^2 - 9$ in the x-axis and so has equation: ✓

$y = -[(x-2)^2 - 9]$

$\Rightarrow y = -(x-2)^2 + 9$

$\Rightarrow y = 9 - (x-2)^2$ ✓

2 marks

Reflection in *x*-axis

- If graph $y =$ (expression in x) is reflected in the x-axis then the equation of the image or new graph is given by $y = -$(expression in x). In this case you require the negative of $(x - 2)^2 - 9$.

Equation

- It is better if you simplify $-[(x - 2)^2 - 9]$. In a simpler example $-(a - b) = -a + b$. Following this pattern gives $-(x - 2)^2 + 9$.
- Notice that, for example, $-2 + 3 = 3 - 2$ so $-(x - 2)^2 + 9$ can be written $9 - (x - 2)^2$. This is the more usual form for this expression.

SUCCESS GUIDE: pages 71, 75

Q5. (a) $\overrightarrow{PA} = \overrightarrow{PD} + \overrightarrow{DA}$
represents $\boldsymbol{v} + (-\boldsymbol{u}) = \boldsymbol{v} - \boldsymbol{u}$ ✓

1 mark

Expression
- $\overrightarrow{AD}$ represents $\boldsymbol{u}$. Then $\overrightarrow{DA}$ represents $-\boldsymbol{u}$.

Q5. (b) $\overrightarrow{BD} = \overrightarrow{BP} + \overrightarrow{PD}$
now $\overrightarrow{BP} = \overrightarrow{PA}$, which represents
$\boldsymbol{v} - \boldsymbol{u}$ from part (a)
So $\overrightarrow{BD}$ represents
$\boldsymbol{v} - \boldsymbol{u} + \boldsymbol{v} = 2\boldsymbol{v} - \boldsymbol{u}$ ✓
($\overrightarrow{BP}$)($\overrightarrow{PD}$)

1 mark

Expression
- Two directed line segments like $\overrightarrow{PA}$ and $\overrightarrow{BP}$ with the same length and direction will represent the same vector.

SUCCESS GUIDE: pages 112, 113

Q6. $\dfrac{2-a}{a^2} + \dfrac{1}{a}$ ✓

$= \dfrac{2-a}{a^2} + \dfrac{a}{a^2}$ ✓

$= \dfrac{2-a+a}{a^2} = \dfrac{2}{a^2}$ ✓

3 marks

Common denominator
- To add, for example, $\frac{1}{2} + \frac{1}{3}$ you need to change the fractions to $\frac{3}{6} + \frac{2}{6}$ so that they have a 'common denominator' of 6.
  In this case $\frac{2-a}{a^2} + \frac{1}{a}$ cannot be added until their denominators are the same. 'a^2' is the simplest such 'common denominator'.

Equivalent fraction
- $\frac{1}{a} = \frac{1 \times a}{a \times a} = \frac{a}{a^2}$. You are always allowed to multiply (or divide) the top and bottom of a fraction by the same number or expression.

Simplify
- $\frac{3}{6} + \frac{2}{6} = \frac{3+2}{6}$: the numerators are added once the denominators are the same. In this case the numerators are $2 - a$ and a giving $2 - a + a = 2$.

SUCCESS GUIDE: page 33

Q7. $P = \sqrt{9 - Q^2}$

$\Rightarrow P^2 = 9 - Q^2$ ✓

$\Rightarrow Q^2 = 9 - P^2$ ✓

$\Rightarrow Q = \sqrt{9 - P^2}$ ✓

3 marks

1st step
- Your aim is to isolate the letter Q. 1st step is to remove the square root. This is done by squaring both sides.

2nd step
- Add Q^2 to both sides and substract P^2 from both sides. This isolates Q^2.

Final step
- To remove the 'squaring' you must 'square root' both sides.
- Technically you should write $\pm\sqrt{9 - P^2}$ but will not be penalised for not doing this!

SUCCESS GUIDE: pages 64–66

Q8. For x-axis intercepts set $y = 0$

So $5 \sin x° - 1 = 0$ ✓

$\Rightarrow 5 \sin x° = 1$

$\Rightarrow \sin x° = \dfrac{1}{5}$ ✓

Since this is positive $x°$ is in the 1st and 2nd quadrants

The 1st quadrant angle is given by

$\sin^{-1}\left(\dfrac{1}{5}\right) = \sin^{-1}(0{\cdot}2) = 11{\cdot}5°$ ✓

The 2nd quadrant angle is

$180° - 11{\cdot}5° = 168{\cdot}5°$

So the required x-coordinate values are $11{\cdot}5$ and $168{\cdot}5$. ✓

4 marks

Equation
- This mark is for obtaining the correct equation.

Rearrangement
- The correct rearrangement of the equation gives you a value for cos $x°$.

***x*-intercept**
- Check that 11·5 is 'sensible' by looking at the graph.

***x*-intercept**
- use  as a memory aid:

sin $x°$ is positive in the 1st *and* 2nd quadrants.

SUCCESS GUIDE: pages 96, 97

Q9.

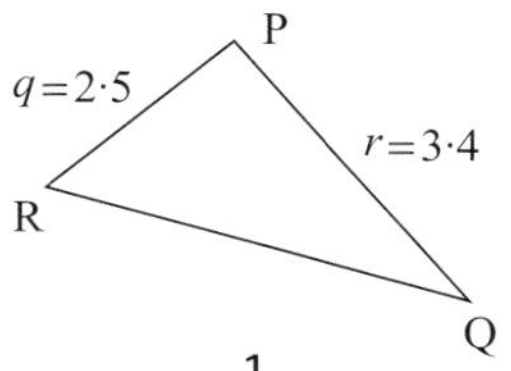

Area of triangle

$= \dfrac{1}{2}qr \sin \mathrm{P}$ ✓

so $4{\cdot}2 = \dfrac{1}{2} \times 2{\cdot}5 \times 3{\cdot}4 \times \sin \mathrm{P}$

$\Rightarrow 4{\cdot}2 = 4{\cdot}25 \times \sin \mathrm{P}$

$\Rightarrow \dfrac{4{\cdot}2}{4{\cdot}25} = \sin \mathrm{P}$

so $\sin \mathrm{P} = 0{\cdot}9882...$ ✓

$\Rightarrow$ angle $\mathrm{P} = \sin^{-1}(0{\cdot}9882...)$

$= 81{\cdot}2...°$

so angle $\mathrm{P} \doteqdot 81°$ ✓

(to 2 significant figures)

3 marks

Strategy
- Normally the formula 'Area $= \dfrac{1}{2}ab \sin C$' is used to find the area but in this case the area is given so an equation is set up and solved first for sin P.

Calculation
- The aim is to find a value for sin P.

Angle
- Use [sin⁻¹] [ans] [EXE] on your calculator where 'ans' uses the previous answer, 0·9882..., on the calculator display.
- [EXE] and [=] are the same – it depends on your calculator.

SUCCESS GUIDE: pages 96, 104, 105

Q10. (a) Set $y = 0$ for x-axis intercepts

so $k(x - p)(x - q) = 0$

$\Rightarrow x - p = 0$ or $x - q = 0$

$\Rightarrow x = p$ or $x = q$

From the given graph this gives: ✓

$p = -2$ and $q = 6$ ✓

2 marks

Intercepts
- There is a correspondence between factors and x-axis intercepts:

Factor		Intercept
$x - p$	⟷	$(p, 0)$
$x - q$	⟷	$(q, 0)$

- Since $p = -2$ the factor $x - p = x - (-2) = x + 2$ so $x + 2$ as a factor gives $(-2, 0)$ as an x-axis intercept.

SUCCESS GUIDE: page 72

Answers to Exam C

Exam C: Paper 1

Q1. A (k, k^2) B $(1, k)$

So $m_{AB} = \dfrac{k^2 - k}{k - 1}$ ✓

$= \dfrac{k(k-1)}{k-1} = k$ ✓ ✓

3 marks

Gradient
- Use the formula

$m = \dfrac{y_2 - y_1}{x_2 - x_1}$

SUCCESS GUIDE: page 36.

Factorisation
- k is a common factor.

Cancellation
- You should make clear in your solution that the factor $k - 1$ cancels.

SUCCESS GUIDE: pages 29, 30.

Q2. $\dfrac{2}{3} \div 1\dfrac{1}{3}$

$= \dfrac{2}{3} \div \dfrac{4}{3} = \dfrac{\frac{2}{3}(\times 3)}{\frac{4}{3}(\times 3)}$ ✓

$= \dfrac{2}{4} = \dfrac{1}{2}$ ✓

2 marks

Division of Fractions
- Change mixed fractions to 'top-heavy'
- An alternative method to the one given is to 'invert then multiply'.

So $\dfrac{2}{3} \div \dfrac{4}{3} = \dfrac{2}{3} \times \dfrac{3}{4} = \dfrac{6}{12} = \dfrac{1}{2}$

Calculation
- Remember to 'cancel down' if possible.

SUCCESS GUIDE: page 122

Q3. $f(x) = x(2 - x)$

$\Rightarrow f(-1) = -1 \times (2 - (-1))$ ✓

$= -1 \times (2 + 1)$

$= -1 \times 3 = -3$ ✓

2 marks

Substitution
- for $f(-1)$, every occurrence of x is replaced by the value -1. No further calculation is needed to gain the 1st mark.

Calculation
- The 2nd mark here is for the subsequent calculation after the substitution.
- Remember: brackets first and that subtracting a negative is the same as adding the positive value.

SUCCESS GUIDE: page 47

Q4. $3 + 2x < 4(x + 1)$

$\Rightarrow 3 + 2x < 4x + 4$ ✓

(now subtract 4 from each side)

$\Rightarrow -1 + 2x < 4x$

(now subtract $2x$ from each side)

$\Rightarrow -1 < 2x$ ✓

$\Rightarrow \frac{-1}{2} < x$

so $x > -\frac{1}{2}$ ✓

3 marks

Brackets
- Both terms are multiplied by 4. A common mistake here would be $4x + 1$.

Simplification
- The 'balancing process' aims to get letters on one side of the inequality sign and numbers on the other side.
- Best to avoid $-2x < 1$. The next step here would be divide by -2 and 'swap the sign round': $x > \frac{1}{-2}$

Answer
- Notice if $a < b$ then $b > a$

SUCCESS GUIDE: page 50

Q5. (a) $9y^2 - 4$

$= (3y)^2 - 2^2$

$= (3y - 2)(3y + 2)$ ✓

1 marks

Difference of squares
- The pattern is: $A^2 - B^2 = (A - B)(A + B)$.
- You should always check that your answer multiplies out giving the original expression (use 'FOIL').

SUCCESS GUIDE: page 27

Q5. (b) $\frac{9y^2 - 4}{15y - 10}$

$= \frac{(3y-2)(3y+2)}{5(3y-2)}$ ✓

$= \frac{3y+2}{5}$ ✓

2 marks

Factorising
- When simplifying algebraic fractions the 1st step is to factorise both expressions.
- 'Hence' appearing in a question means you should use the answer that you got in the previous part of the question: in this case $9y^2 - 4 = (3y - 2)(3y + 2)$.

Cancelling
- Any factor, e.g. $3y - 2$ that appears on the top and the bottom can be cancelled.

SUCCESS GUIDE: pages 29, 30.

Q6. $P = \frac{W}{4A}$

(×A) (×A)

$\Rightarrow PA = \frac{W}{4}$ ✓

(÷P) (÷P)

$\Rightarrow A = \frac{W}{4P}$ ✓

2 marks

1st rearrangement
- The subject of the formula, A, in this case appears on the bottom of the fraction i.e. W is divided by A. The inverse process is multiplication so multiply both side of the formula by A.

2nd rearrangement
- A is multiplied by P. The inverse process is division so divide both sides by P.

SUCCESS GUIDE: page 64

Q7. (a)

x	$x-\overline{x}$	$(x-\overline{x})^2$
503	2	4
504	3	9
497	−4	16
495	−6	36
506	5	25

$\Sigma(x-\overline{x})^2 = 90$ ✓

$$S = \sqrt{\frac{\Sigma(x-\overline{x})^2}{n-1}} = \sqrt{\frac{90}{4}}$$ ✓

$= 4{\cdot}743... \doteqdot 4{\cdot}7$ ✓

(to 1 decimal place)

3 marks

Squared Deviations

- This mark is gained for correctly calculating the values in the last column of the table, i.e. $(x-\overline{x})^2$.
- No negative values should ever appear in this column. Remember when you square a quantity your answer will be positive or zero
- The mean, $\overline{x}$, has been given to you in the question and so you do not need to calculate it again!

Substitution

- This mark is for correct substitution of the values $\Sigma(x-\overline{x})^2 = 90$ and $n - 1 = 4$ into the standard deviation formula.

Calculation

- Remember always that there is a square root to be taken at the end of a standard deviation calculation.
- Any reasonable rounding is acceptable
- Using the other formula gives $\sum x = 2505$, $\sum x^2 = 1\,255\,095$ and $n = 5$ resulting in $s \doteqdot 4{\cdot}7$

SUCCESS GUIDE: pages 130, 131

Q7. (b) Yes it did. The new standard deviation of 3·5 is less than the previous value of 4·7 so there was less variation about the mean. ✓

1 mark

Statement

- Your explanation must use the standard deviation statistics that you calculated in part (a) and are also given in the question. The greater the standard deviation the greater the variation about the mean (and vice versa).

Q8. (a)

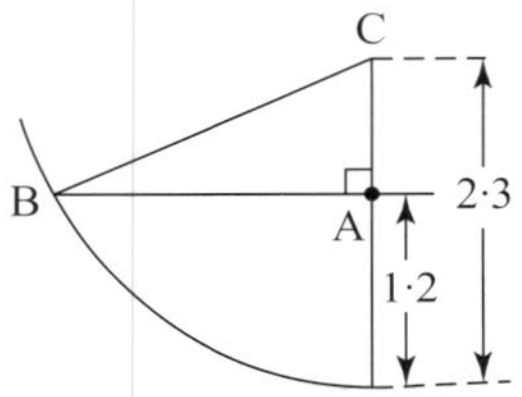

The radius of the fuselage is 2·3 m
AC = 2·3 − 1·2
= 1·1 m (see diagram) ✓

Use Pythagoras' theorem in triangle ABC:

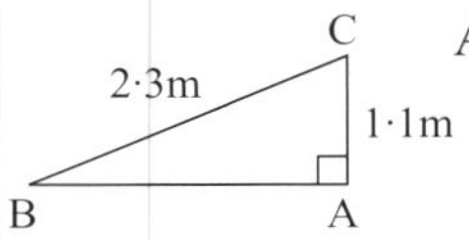

$AB^2 = BC^2 - AC^2$ ✓

$= 2{\cdot}3^2 - 1{\cdot}1^2$

$= 4{\cdot}08$ ✓

so $AB = \sqrt{4{\cdot}08} = 2{\cdot}019...$

By symmetry: width of floor
= 2 × AB

so $x = 2 \times 2{\cdot}019... = 4{\cdot}039...$ ✓

$\Rightarrow x \doteqdot 4{\cdot}04$ metres.

4 marks

Strategy

- Triangle ABC is right-angled and therefore Pythagoras' theorem can be used.

Pythagoras' theorem

- Notice that CB is a radius and that CA is part of a radius. Any radius in this circle has a length $\frac{1}{2} \times 4{\cdot}6 = 2{\cdot}3$ m

Calculation

- AB is one of the smaller sides in triangle ABC so the calculation involves a subtraction
- Remember that rounding off decimal answers should only be done at the end of the problem when the final answer is given.

Solution

- By symmetry AB is half the width of the compartment floor.

Q8. (b) By symmetry (see diagram) the ceiling is as far above C as the floor is below C.

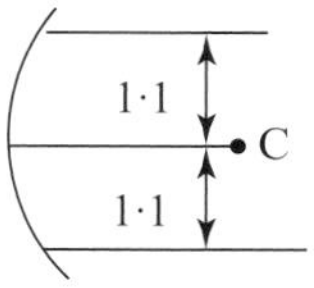

Height $= 2 \times 1{\cdot}1$
$= 2{\cdot}2$ metres ✓

1 mark

Solution

- Two equal and parallel chords in a circle will be the same distance from the centre because the diagram is symmetric.

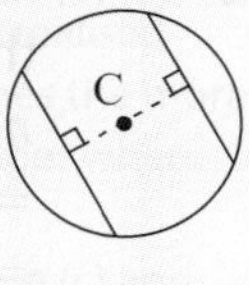

SUCCESS GUIDE: pages 76, 85

Q9. For the points of intersection solve:

$\cos x° = 0{\cdot}2$ ✓

$x°$ is in the 1st or 4th quadrants

1st quadrant angle is given by:

$x = \cos^{-1}(0{\cdot}2) \doteqdot 78{\cdot}5$ ✓

4th quadrant angle is given by:

$x = 360 - 78{\cdot}5 = 281{\cdot}5$

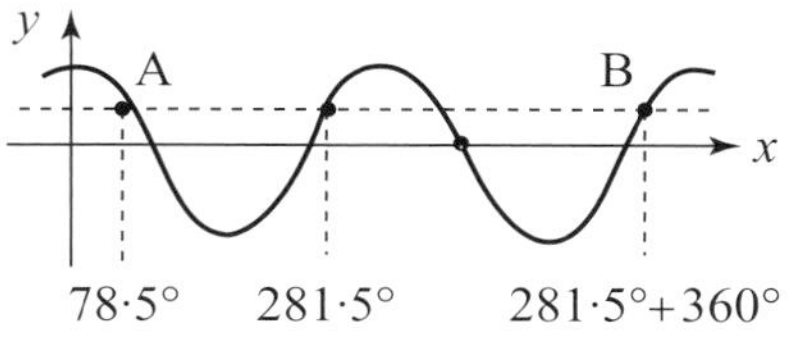

The x-coordinate of A is 78·5

From the graph above

B lies 1 cycle on from the 4th quadrant intersection

The x-coordinate of B is $281{\cdot}5 + 360 = 641{\cdot}5$ ✓

3 marks

Strategy

- At each of the points of intersection the y-coordinate is 0·2 and the x-coordinate satisfies $\cos x° = 0{\cdot}2$.
- The appearance of the equation $\cos x° = 0{\cdot}2$ and an attempt to solve it will gain you this mark.

1st quadrant angle

- Using $\boxed{\cos^{-1}}$ on your calculator will always give you the 1st quadrant angle (don't use this key with a negative value).

Other angles

- You should identify the 4 quadrants on the graph that is given:

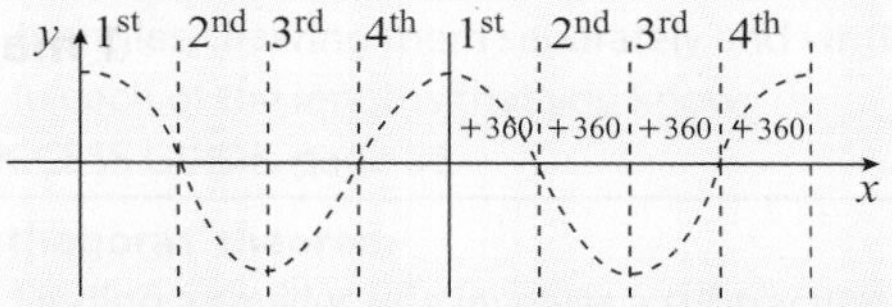

It should then be clear that the x-coordinate of B is given by the 4th quadrant value + 360°

- The 4th quadrant value is obtained using this diagram:

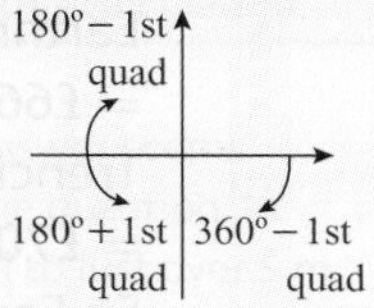

SUCCESS GUIDE: pages 96, 97

Q10. (a)

4th line: $8 + 1 = 4 \times 6 - 5 \times 3$ ✓

1 mark

Pattern extension

- The order of writing the line cannot be altered.

Q10. (b)

1st line:	2	+	1	=	1	×	3	−	2	×	0	
2nd line:	4	+	1	=	2	×	4	−	3	×	1	
3rd line:	6	+	1	=	3	×	5	−	4	×	2	
4th line:	8	+	1	=	4	×	6	−	5	×	3	
⋮	⋮		⋮		⋮		⋮		⋮		⋮	✓

nth line: $2n + 1 = n\,(n+2) - (n+1)\,(n-1)$ ✓

2 marks

1st expression

- For finding one of $2n + 1$ or $n(n+2)$ or $(n+1)\,(n-1)$ you will gain the 1st mark here.

Remaining expressions

- Careful use of brackets is essential:
  $n \times n + 2\ (= n^2 + 2)$ is not the same as $n(n+2)$
  $n + 1 \times n - 1\ (= 2n - 1)$ is not the same as $(n+1)\,(n-1)$
- Always check: e.g. substituting $n = 3$ should produce the 3rd line.